SCIENCE INVESTIGATORS

Investigating ENERGY Through Modeling

Derek Miller

Cavendish Square
New York

Published in 2020 by Cavendish Square Publishing, LLC
243 5th Avenue, Suite 136, New York, NY 10016

First Edition

Library of Congress Cataloging-in-Publication Data

Names: Miller, Derek L., author.
Title: Investigating energy through modeling / Derek Miller.
Description: First edition. | New York : Cavendish Square, [2020] | Series: Science investigators | Audience: Grades 5 to 8. | Includes bibliographical references and index.
Identifiers: LCCN 2019012327 (print) | LCCN 2019013607 (ebook) | ISBN 9781502652423 (ebook) | ISBN 9781502652416 (library bound) | ISBN 9781502652409 (pbk.)
Subjects: LCSH: Force and energy--Experiments--Juvenile literature. | Science--Methodology--Juvenile literature.
Classification: LCC QC73.4 (ebook) | LCC QC73.4 .M55 2020 (print) | DDC 531/.6078--dc23
LC record available at https://lccn.loc.gov/2019012327

Editor: Caitlyn Miller
Copy Editor: Nathan Heidelberger
Associate Art Director: Alan Sliwinski
Designer: Ginny Kemmerer
Production Coordinator: Karol Szymczuk
Photo Research: J8 Media

The photographs in this book are used by permission and through the courtesy of:
Cover ©iStockphoto.com/SerrNovik; p. 4 FooTToo/Shutterstock.com; p. 7 John Panella/Shutterstock.com; p. 10 Serjio74/Shutterstock.com; p. 12 Pixel_Pig/E+/Getty Images; p. 13 Racheal Grazias/Shutterstock.com; p. 15 Volodymyr Maksymchuk/Shutterstock.com; p. 16 Sean Murphy/The Image Bank/Getty Images; p. 17 Busurmanov/Shutterstock.com; p. 20 Rob Lewine/Getty Images; p. 22 Mary Lynn Strand/Shutterstock.com; p. 24 Anan Kaewkhammul/Shutterstock.com; p. 26 imammoto/Shutterstock.com; p. 27 NoPainNoGain/Shutterstock.com; p. 29 snapgalleria/Shutterstock.com; p. 30 GraAssimo/iStock/Getty Images; p. 32 Catalin Petolea/Shutterstock.com; p. 34 Leezsnow/E+/Getty Images; p. 35 Joseph Sohm/Shutterstock.com; p. 37 PitchyPix/Shutterstock.com; p. 41 Noradoa/Shutterstock.com.

Printed in the United States of America

Contents

Electric lights, like those shown here, are just one way humans use energy.

CHAPTER 1

What Is Energy?

Everything contains energy. Energy comes in many different forms, including motion, sound, light, heat, and electricity. People harness these types of energy to do amazing things. From powering smartphones to exploring space, energy makes it all possible.

MOTION ENERGY

The scientific term for the energy of something in motion is kinetic energy. When scientists invent new ways to move people and goods faster, they work with kinetic energy.

At the simplest level, cars use gasoline to create kinetic energy. The energy stored in fuel is used to make the car accelerate, or go faster. The kinetic energy of the moving car increases as it accelerates. When it is moving fast, it has a huge amount of kinetic energy.

The kinetic energy of a moving car decreases when the driver brakes. Energy never really goes away. It just changes from one kind of energy to another. A braking car loses kinetic energy because the energy is converted into heat. Most of the heat is transferred to the brakes. In fact, the brakes on a car can heat up so much that they turn red.

Temperature is related to kinetic energy. All matter is made up of particles that are too small to see. The particles are always moving around. The faster they are moving, the hotter an object is. The slower they are moving, the colder an object is. The energy given off by these moving particles is called thermal energy. We experience it as heat.

TRANSFERRING ENERGY

Thermal energy flows naturally between two objects that touch. When ice is dropped into hot water, the ice melts, and the water cools off.

A bolt of lightning releases a huge amount of energy.

Lightning is an example of a natural phenomenon that transfers huge amounts of energy. When lightning occurs, a storm's energy transfer takes many different forms. Thunder is the release of energy through sound. Light energy is visible in the bolt of lightning. At the location of the lightning, thermal energy and electricity are also released. The ground can be scorched black from the lightning strike.

Natural phenomena are not the only time that energy is transferred. A great deal of science and engineering has to do with man-made methods of transferring energy. Without electricity, life today would be very different. The things we take for granted that make our lives easy—like air conditioners and refrigerators—would not exist. All these devices require electricity.

To make electricity, other types of energy are converted. Wind turbines convert kinetic energy from moving air into electricity. Power plants that burn coal convert thermal energy to electricity. Engineers can't make electricity out of nothing. They must find energy in other forms and convert it.

USING MODELS TO UNDERSTAND AND EXPERIMENT

Scientists often make models to understand how the world works. A model can be made from physical objects, like a diorama you make for a school project. Models can also be mathematical equations, computer simulations, and charts and graphs. Scientists use models to test hypotheses and to show their ideas to others. In this book, you will have the chance to make models that show how energy shapes life on Earth.

Early Models of Motion

Today, we talk about "kinetic energy." In the past, scientists called it "motion." Scientists and mathematicians debated what mathematical equation best described motion. Seventeenth-century philosopher René Descartes thought mass times velocity added up to "quantity of motion." (Mass is related to weight, and velocity is like speed.) According to Descartes, mass and velocity had equal importance to energy. Years later, another mathematician, Gottfried Leibniz, disagreed with Descartes. He argued that the equation should read mass times velocity squared. According to Leibniz, velocity generated more energy than mass.

An experiment was done to see which equation was better. Willem 's Gravesande dropped brass balls onto a soft clay block. He found that a ball moving twice as fast left an indentation four times as deep. Two squared is four. A ball moving three times as fast left an indentation nine times as deep. Three squared is nine. This proved Leibniz right. The better formula was energy equals mass times velocity squared. Velocity mattered more than mass.

In the end, Leibniz improved Descartes's equation. Descartes's model was wrong, but it was an important step in understanding energy. Part of making a model is understanding that it can be improved later. Improving a model that already exists can be just as important as making a new model.

Kinetic energy makes it difficult for a moving train to stop quickly.

CHAPTER 2

Modeling Kinetic and Potential Energy

Energy cannot be created or destroyed. It just changes its form. When a car brakes, the changing energy heats up the brakes. When we listen to music, electrical energy is converted into sound energy. Still, there are some examples that aren't so clear cut. Imagine throwing a ball into the air. At its highest point, it isn't moving at all. Where did the ball's starting kinetic energy go? Unlike a car, there isn't much heat produced. The answer is something called potential energy.

Swinging is an example of the ways potential energy and kinetic energy are related.

POTENTIAL ENERGY

Potential energy exists in a system of objects. In the example of a ball thrown into the air, the objects are the ball and Earth. When the ball is in the air, its potential energy is due to gravity. Earth's gravity pulls the ball down. After the ball's upward kinetic energy is exhausted, the ball accelerates toward the ground.

One way to think about potential energy is that objects store it—often based on their position. In fact, potential energy is sometimes called stored energy. Imagine a car driving up a steep hill. It takes a lot of energy for it to go forward. The driver has to give it a lot of gas, and the engine works hard. Once

Roller coasters convert potential energy into kinetic energy when the cars drop toward the ground.

the car reaches the top, however, it has potential energy. The driver can stop pressing on the gas pedal, and the car will just roll down the hill. The car's position on the hill means it has stored energy.

USING POTENTIAL ENERGY

Potential energy has many uses in the real world. Take the example of a roller coaster. Roller coasters usually begin by pulling a train full of people to a high point on the tracks. In this way, the potential energy of the train increases. As the train goes downhill, the potential energy is converted into kinetic energy. The passengers accelerate toward the ground due to gravity.

The train may start going back uphill, or even do a loop. Going uphill, the kinetic energy is converted back into potential energy as the ride slows. At the end, the train full of people brakes and comes to a stop. Braking converts the last of the kinetic and potential energy into heat.

Many famous examples of potential energy rely on gravity. An object's position relative to Earth is a straightforward way to see how energy can be stored. Objects that are high up can fall or roll downward due to gravity. However, gravity is not the only source of potential energy. Magnets are another example of potential energy. One magnet doesn't do much, but when you get it close to another magnet, its stored energy can be seen. The two magnets will attract or repel each other.

MAKE YOUR OWN MODEL OF POTENTIAL ENERGY

You can investigate potential energy on your own. All you need to do is find a wheeled vehicle—like a toy car—and something to make a ramp out of. A ramp is useful because it lets you see gravity work in slow motion. Objects fall through the air quite quickly, but they roll down a ramp slower.

You can make your own model with a toy car to show how potential energy works.

Use objects you find around the house or outside to make your ramp.

First, set up a ramp. You will need a flat object that your car can roll over. Ideally, it will be hard so that the wheels turn easily. A wide, wooden board or sheet of plywood will work. If you don't have something hard, you can try a couch cushion or some other soft object. If you're lucky, you might even have a toy in your house that includes a slope and a wheeled vehicle—train tracks or a racing track will work well for this model.

Make sure your ramp has a flat surface at its end that your car can roll across.

Once you have your ramp set up, try putting your car at the top and releasing it. Hopefully, it rolled down the ramp to the floor and kept going. If it fell off the ramp or didn't roll smoothly onto the floor, try to fix the problem. You might need a wider ramp if it fell off the side. Otherwise, try releasing it closer to the bottom. If there is a gap at the bottom of the ramp, see if you can think of a way to fix the problem. You might be able to find an object that bridges the gap from the bottom of the ramp to the floor.

Often, scientists need to troubleshoot problems when they are working with models and experimenting. They must think of ways to test just one issue that they want to measure. Any problems with their experiment that introduce other issues need to be fixed.

Before you test your model, remember that potential energy and kinetic energy are related. An object that has more potential energy stored will have more kinetic energy when it is released. Kinetic energy is proportional to the mass and velocity of an object. The mass of the car stays the same, but its velocity changes. The faster it moves, the more kinetic energy it has.

Now, try putting the car four car lengths up the ramp and releasing it. It should roll down the ramp and eventually come to a stop. Put a piece of tape or paper down where it stopped rolling. Write down a hypothesis about what will happen if you put the car two car lengths up the ramp and release it.

Mark the ground where your car stopped rolling.

Next, test your hypothesis. Mark where the car stopped rolling after being released from two car lengths up. Was your hypothesis correct?

Test Yourself

1. Did the car have more kinetic energy the first time—released higher up the ramp—or the second time—released closer to the bottom?
2. Does the car have more or less potential energy the higher it is on the ramp?
3. Did you notice any other kinds of energy besides kinetic and potential energy?
4. When does the car lose all its potential energy?

Answers:

1. The car had more kinetic energy the first time. It should have rolled farther because it had more velocity.
2. The car has more potential energy the higher it is. You can see this with your own eyes because it rolls faster and farther the higher it is placed on the ramp.
3. Sound energy was also present as the car rolled and maybe crashed. The wheels of the car may also have heated up.
4. As soon as the car leaves the ramp, it has no more potential energy. On the flat surface below the ramp, the car only has kinetic energy until it comes to a stop.

Eating balanced meals gives people plenty of energy.

CHAPTER 3

Modeling Energy in Living Things

All living things need energy. People get energy from the food they eat. The energy in food lets humans move around and keeps them alive. Other animals also get their energy from food, although their diets may be very different from people's diets. The energy that people and animals get from food starts as energy from the sun. Animals can't use sunlight to stay alive, but plants convert energy from the sun into energy that animals use when they eat foods like vegetables, fruits, and grasses.

PHOTOSYNTHESIS AND ENERGY

Plants use sunlight in a process called photosynthesis. Photosynthesis is a chemical reaction. It takes place in the leaves of plants. During photosynthesis, a plant needs sunlight, water, and carbon dioxide (a gas in air). From these three things, plants can create oxygen (another gas released into the air) and sugars. These sugars are what plants use for energy. The sugars power the growth of the plant and keep it alive.

Almost all energy that people and animals use begins as sugars that plants create during photosynthesis. When people

Plants get their energy from the sun.

and animals eat plants, they use these sugars as energy. Animals that only eat meat—called carnivores—get energy from plants in a roundabout way. They eat the meat of animals that have themselves gotten energy from plants. Sometimes, the chain of energy transfer can be quite long. A carnivore may eat other carnivores, but at some point, an animal ate the plants that provide the energy for all the carnivores in the chain.

ECOSYSTEMS

The plants and animals in a given place and the complex relationships between them are called an ecosystem. When scientists look at an ecosystem, they figure out how energy and matter is transferred between all the organisms—or living things—in the ecosystem. Plants play the important role of turning matter that is not food into matter that is food. Food provides animals and humans with both energy and matter. The matter is used to grow and also to heal when injured. The energy lets animals move and keep warm. Plants take in energy and matter that animals cannot use for food—including carbon dioxide and sunlight—and make sugars that animals can use for food.

Plants get most of the matter they need from the air in the form of carbon dioxide and through water from the ground.

An ecosystem is made up of all the plants and animals that live in a place.

However, plants do get some matter from soil. Plants need some nutrients from the ground to grow. Nutrients like calcium and potassium exist in the ground. Animals get these same nutrients by eating plants or animals that have eaten plants.

When animals die, these nutrients make their way back into the soil through decomposition. Decomposition is the process of matter from dead animals going back into the ground. The body of the animal breaks down and provides matter for plants. In this way, the matter in an ecosystem is recycled.

Decomposition is driven by organisms called decomposers. Examples of decomposers are fungi and earthworms. They feed on dead animals and break them down. Then, plants can once again use the nutrients from the dead animals' bodies to create food for themselves and for other animals.

MAKE YOUR OWN MODEL: A FOOD CHAIN

Scientists use different models to show how energy and matter move through an ecosystem. One model is called a food chain. A food chain illustrates which organisms eat other organisms in an ecosystem. To make your own food chain, all you need is a pencil and a piece of paper. The first step of a food chain is always the sun, so you should start with a picture of the sun. Then, plants get energy and matter from the sun, water, and air. Draw a picture of grass on your piece of paper.

You can look at this drawing of a grasshopper as you make your own.

Next, try to think of an animal that eats grass. For this example, we'll use a grasshopper. Draw a grasshopper to the right of the grass and then draw an arrow pointing from the grass to the grasshopper. The arrow represents the transfer of energy and matter from the grass to the grasshopper.

The next animal you draw needs to eat grasshoppers. Can you think of one? Birds, snakes, and some mice all eat grasshoppers. Draw one of these animals and connect it to the grasshopper with an arrow. For the last step, think of an animal that eats whichever animal you picked. One likely candidate is a fox. Foxes eat birds, snakes, and mice, as well as many other small animals. After you draw a fox or other carnivore, you should

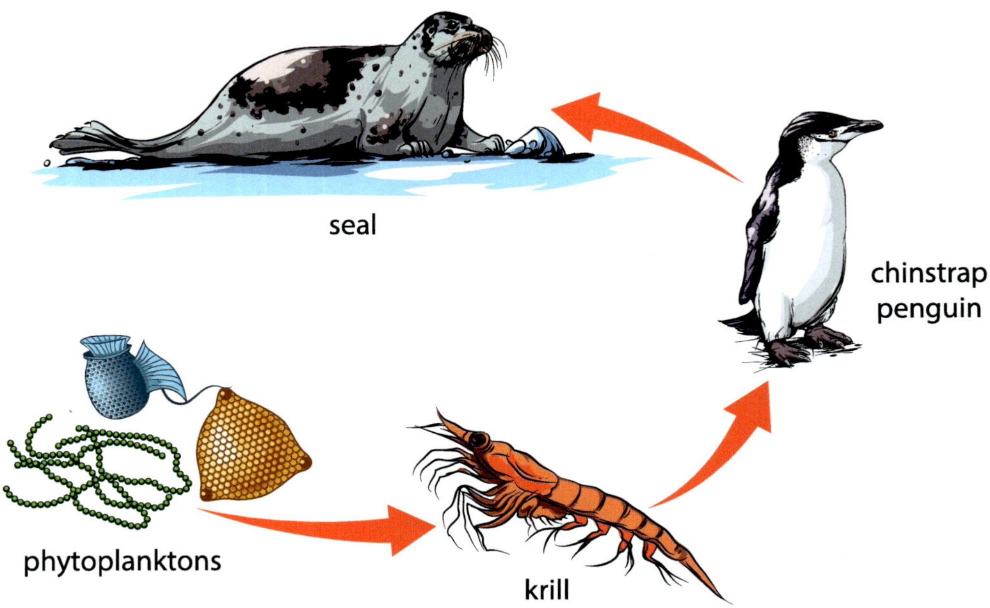

How does this Antarctic food chain compare to your food chain?

have a food chain with four organisms on it. Your food chain illustrates how energy from the sun is captured by plants and then powers the entire ecosystem through food.

MAKE YOUR OWN FOOD WEB

A food chain is fairly simple. Each organism has at most one arrow pointing to it and one arrow pointing away from it. The last carnivore in the chain usually doesn't have an arrow pointing away from it since it is the apex predator. Other animals do not eat apex predators because apex predators are the biggest or strongest in their ecosystem.

A food web is like a food chain, but it is more complex. In a food web, you can show more complicated relationships between organisms because you can link them with any number of arrows. To start your own food web, choose a couple different plants, such as grass, blackberry bushes, and oak trees. Many different animals eat these plants. Try to think of two animals that eat each one. For trees, you can include animals that eat the nuts or seeds the trees produce, like squirrels and chipmunks. Some of these animals might even eat two or all three of the plants. You can draw arrows from multiple plants to these animals.

Then, add some predators into your food web. They might eat one or all of the animals you've drawn. Some predators

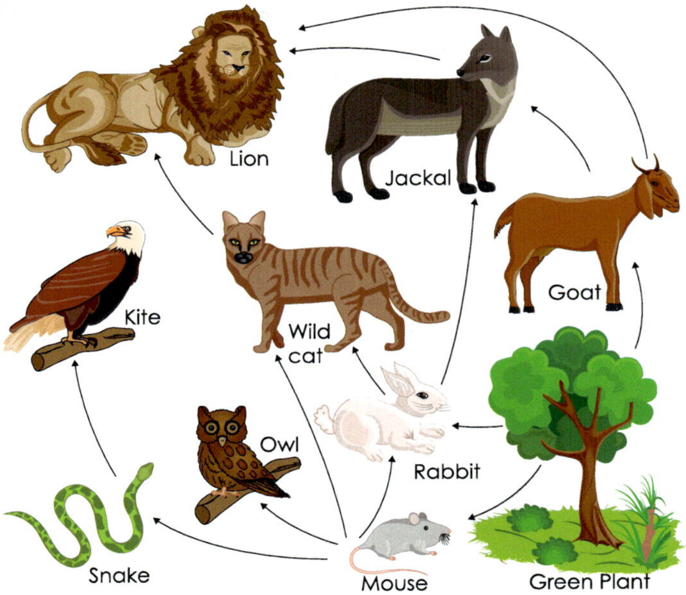

Biology **Food Web**

An animal or plant can have many arrows connected to it in a food web.

also eat plants. For example, bears eat berries and nuts as well as meat and fish. If you include a bear in your food web, it will have a lot of arrows going to it. Try to research what the animals you draw eat, and make sure you include all the arrows.

Once you've finished, you can see how complex an ecosystem is. Ecosystems are delicate. When humans introduce a new animal into one, it can have devastating consequences. Rats and cats are two animals that humans have spread around the globe. In some places, these animals went on to kill entire species, especially birds. Looking at your food web, you can see how the loss of some species could affect many other species.

Test Yourself

1. How does the sun fit into your food web? Draw the sun and add arrows showing how energy from sunlight fuels some organisms directly (but not others).
2. Where do decomposers belong on your food web? Add an earthworm—a decomposer—to your food web. Add arrows as well. (Remember that earthworms eat dead plants and animals.)
3. How do the nutrients produced by earthworms affect your food web? Nutrients produced by earthworms go back into the soil and are used by some organisms as matter. Draw arrows to the correct organisms.

Answers:

1. The sun should have arrows going from it to all the plants.
2. All the plants and animals should have arrows going from them to the earthworm. Some animals, like birds, also eat earthworms.
3. Earthworms should have arrows going from them to plants. While plants don't consume earthworms, they need the nutrients that decomposers leave in the soil.

It takes a huge amount of energy for a plane to fly.

CHAPTER 4

Modeling Renewable Energy

Humans get the energy to power their bodies from eating food, but food is only a small part of the energy that people use. Energy also powers human civilization. Electricity heats and cools houses, making them livable in extreme climates. Cars, trucks, buses, airplanes, and trains use energy to move people—and goods—around the world. Without huge amounts of energy, modern life would not be possible.

FOSSIL FUELS AND RENEWABLE ENERGY

Fossil fuels provide most of the energy that humans use. Coal, crude oil, and natural gas are all fossil fuels. Products made from crude oil, like gasoline, power cars and other vehicles. Coal and natural gas are used to produce electricity.

Fossil fuels are created naturally. Dead plants and animals are buried in Earth's crust—the outer layer of the planet. Over hundreds of millions of years, heat and pressure turn the remains of plants and animals into fossil fuels.

Oil wells bring crude oil from inside Earth to the surface.

Scientists divide sources of energy into two categories: renewable and nonrenewable. Renewable energy is harnessed from sources that are not depleted—or used up. Solar and wind power are examples of renewable energy. The sun and wind are not used up by solar panels and wind turbines.

Meanwhile, fossil fuels are nonrenewable. They are being completely used up to create electricity. For instance, coal is burned to create thermal energy, or heat. The heat is used to boil water and produce steam. The steam powers turbines that produce electricity. At the end of the process, there is electricity but no more coal.

PRODUCING ELECTRICITY

As we learned earlier, energy does not disappear. It just changes its form. Likewise, energy cannot be created from nothing. When people produce electricity, they need to capture another kind of energy and convert it to electricity.

A machine called a turbine is commonly used to convert other kinds of energy into electricity. Turbines use kinetic energy to produce electricity. The kinetic energy spins a generator inside the turbine, and this movement generates electricity.

Coal, natural gas, wind, and water all power turbines through kinetic energy, though the exact way they do it varies. Coal

These turbines in a dam create electricity using flowing water.

and natural gas heat steam that is used to spin the turbine. In wind turbines, the blades spin with the wind. In a dam, blades use the energy of flowing water to make electricity.

THE DOWNSIDES OF ENERGY PRODUCTION

All forms of energy production have some negative effects. They hurt the environment in different ways. The exact downsides depend on the source of the energy. Fossil fuels are worse for the environment than most other sources of energy.

Burning fossil fuels pollutes the air and water. Some of this pollution is poisonous. For example, the exhaust from a car is potentially deadly in an enclosed space. In a crowded city, many

Pollution makes the air gray and unhealthy in many cities around the world.

cars can make the air unhealthy to breathe. Other pollution is bad because it is changing Earth's climate. Climate change is partly driven by fossil fuels. Average temperatures on Earth are increasing due to the burning of fossil fuels. Extreme weather events, like droughts, floods, and heat waves, are becoming more common as a result.

Dams do not pollute the environment, but they can harm the local ecosystem. Dams can flood areas around where they are built. They can also block the movement of animals up and down a river. Scientists carefully study where dams will go to try to prevent this damage to the environment.

Wind turbines do little to harm the environment, but even they have some downsides. The spinning blades can kill birds in the air. Some wind turbines are also quite noisy. The noise can confuse and stress nearby wildlife. People living near wind turbines also complain that they are ugly.

Experts work hard to try to lessen the negative effects of energy production. New inventions could be on the horizon that let people use energy without hurting the planet.

MAKE YOUR OWN MODEL: A WIND TURBINE IN ACTION

To see how renewable energy is produced, you can build your own model of a wind turbine. The model won't have the machinery to generate electricity as it spins, but it will show how blades can capture kinetic energy from moving air.

A pinwheel is powered by moving air, just like a wind turbine is.

Materials

scissors

construction paper

ruler

push pin

thin wooden dowel

glue

Instructions

First, you need to cut your construction paper into an 8-inch by 8-inch (20-centimeter by 20-centimeter) square. Then, use a ruler to draw a straight line from one corner to the opposite corner diagonally across the paper. Do the same for the other corners. Your square paper should now have an "X" drawn across it. The point where the two lines cross is the center of the paper: draw a dot there.

Next, cut in from each outside corner along the lines you've drawn. You can either cut in a bit over halfway to the center dot you drew, or you can measure in 3 inches (7.5 cm) from each corner if you want to be more exact. Once you've made the four cuts, you're ready to start gluing.

Lay the piece of paper on a flat surface and rotate it so that it looks like a diamond. One corner should be facing toward you, one to the left, one away from you, and one to the right. Take the bottom half of the corner facing to the left and fold it to the middle dot. Don't crease the paper, just let it fold naturally. Glue the tip of the piece of paper to the middle dot. You might have to hold it there for a bit so that it sticks. Repeat this step with the left side of the corner facing away from you, then the top side of the corner to your right, and finally the right side of the corner facing toward you. You should now have four blades for your wind turbine model.

Lastly, very carefully pierce the center of your paper with the push pin, and then push the pin into the side of the wooden dowel at one end. Don't push the pin into the wood too far or it might stop it from spinning.

Your model of a wind turbine is ready. Try holding it outside if it's a windy day or blowing on it. The paper blades should capture the kinetic energy of the flowing air and spin. This is exactly how a real wind turbine works. The only thing yours is missing is a generator to convert the captured kinetic energy into electricity.

Wind is a clean, renewable source of energy.

Test Yourself

1. Try blowing on your model wind turbine from different sides. Does it affect how quickly it spins?
2. Why do you think wind turbines reach so high into the sky?
3. See how fast you can get the blades of your wind turbine to spin. Some real blades weigh more than 12,000 pounds (5,440 kilograms)! Can you make an educated guess about which safety measures engineers build into these massive machines?

Answers:

1. The blades should spin faster or slower depending on what direction the air flows across them. For this reason, the tops of wind turbines can rotate to match the direction of the wind.
2. Wind speeds are faster the higher you go. Faster winds let wind turbines capture more energy.
3. Wind turbines are built with brakes inside of them. During high winds, the brakes are applied to stop the blades from spinning out of control. When the brakes fail, some wind turbine blades have come crashing down.

Glossary

apex predator An animal at the top of the food chain. Normally, no other animal tries to eat it.

carnivore An animal that eats other animals but does not eat plants.

convert To change.

decomposer An organism that breaks down dead plants or animals, such as earthworms or some types of fungi.

ecosystem The environment, plants, and animals in an area.

fossil fuels Fuels made from the remains of dead plants and animals. They include coal, oil, and natural gas.

kinetic energy Energy of motion; the faster an object is moving and the heavier it is, the more kinetic energy it has.

organism An individual living thing; for example, a plant, animal, fungus, or bacterium.

photosynthesis The process that takes place in the leaves of plants that creates food energy in the form of sugars from sunlight, water, and a gas in air.

potential energy The energy stored in an object. Often, potential energy exists because an object is positioned to fall or roll downhill due to gravity.

renewable energy Energy that is not used up when humans harness it; for example, solar and wind energy are renewable.

turbine A machine that produces energy—usually electricity—by spinning.

Further Information

BOOKS

Claybourne, Anna. *Forms of Energy*. Chicago, IL: Capstone Press, 2016.

Hirsch, Rebecca E. *Climate Change and Energy Technology*. Minneapolis, MN: Lerner Publishing Group, 2019.

Nardo, Don. *Science and Sustainable Energy*. Science and Sustainability. San Diego, CA: ReferencePoint Press, 2017.

WEBSITES

Introduction to Potential and Kinetic Energy

https://www.texasgateway.org/resource/potential-and-kinetic-energy-1

The Texas Education Agency provides an interactive look at potential and kinetic energy.

Power Up

https://climatekids.nasa.gov/power-up

At this website, NASA explains renewable energy and provides a fun game that illustrates how wind and solar power work.

VIDEOS

Food Chains Compilation

https://youtu.be/CZhE2p46vJk

Crash Course Kids details how energy from the sun flows through an ecosystem.

Renewable Energy 101

https://youtu.be/T4xKThjcKaE

Student Energy explains what renewable energy is and talks about its most common forms.

Will Fossil Fuels Run Out?

https://youtu.be/jjfs_7kwRks

BBC Science presenter Greg Foot examines the question of when the planet will run out of fossil fuels.

Index

Page numbers in **boldface** refer to images.

About the Author

Derek Miller is an educator from Salisbury, Maryland. He is the author of more than a dozen books, including *Science Investigators: Investigating Weather and Climate Through Modeling* and *Science Investigators: Investigating Space Through Modeling*. In his free time, Miller likes to read and travel with his wife.